UNDERSTANDING DEPRESSION

'Help for the helpers'

A practical guide for any and every body involved

DR. RICK APPLETON

ISBN: 978-0-244-38634-4

The 'Black Dog'!

The intention of this little book is to provide some guidelines as to how to battle against that great and powerful enemy we call depression. Hopefully, it will be of direct help to sufferers, but for them this can seem an extremely lonely struggle, so:

Its primary target is anyone who is trying to help them.

Few illnesses affect only those who are afflicted, and seldom is this truer than in this particular malady. To anyone who has never experienced deep depression personally, considerable effort may be required just to understand the nature of the beast and to get an inkling of what it is like.

In the absence of proper insight, the tendency is to make well-meaning remarks like:

> *"Come on old chap, pull yourself together."*
> *"You must snap out of it."*
> *"Things aren't as bad as they seem"*,

which only make things worse!

They tend to infer that the sufferer hasn't been making any effort whatsoever, or likes being like that.

If asked to explain what they mean by "snap out of it", they will be quite unable to say, or offer some other platitude like "Look on the bright side", which I suppose is a beginning, but not very constructive.

What is needed is understanding and a specific action plan.

My own 'understanding' has been acquired, over many years as a family doctor, both from my patients and from other people who are trying to help them and as such is only my personal experience of a very complex subject.

The fact that other therapists may see it from a different point of view does not invalidate either of our opinions.

The important thing is that for me, and hopefully for you also, the concepts I will be describing point the way to

**a positive and clear-cut technique for
firstly, recognizing the enemy and then
establishing possible means to combat it.**

Part of the problem is a reluctance to talk about it, so;

Let's jump in at the deep end.

'SUICIDE'

How does that word affect you?
Is it a 'gut reaction 'of horror?
One of the 'unmentionables', like
Cancer or *Sexually Transmitted Disease*?

You may be thinking
  "How can you be so unfeeling as to be so blunt about it?"
But wait a minute - whilst gut reactions sometimes turn out to
be sustained by common sense, at other times, on deeper

reflection, they prove to be only part of the truth and, before one can form an intelligent opinion, it's advisable to look a little deeper.

If we were considering a 19th Century Japanese, he and his family might regard suicide as a triumph, and, closer to home, the actual fact of suicide might be a longed-for release and the realization of the deepest desire of the perpetrator.

No – the tragedy is not suicide, but what has led up to this awful conclusion.

Chad Varah, the founder of 'The Samaritans' organization, which is dedicated to helping anyone who is suffering from suicidal thoughts said:

> ***"The tragedy is that a person can feel that***
> ***life has dealt so badly with them, that***
> ***they no longer want to go on with it."***

The natural death of an elderly person, who has reached the logical termination of a good, complete life, may be sad for those who love them, but can't, by any stretch of the imagination be regarded as tragic. Indeed, many old people in these circumstances, almost look forward to dying.

On the other hand, many in these circumstance cling to life with the most amazing tenacity, which makes it even more difficult to understand what it is like for a young, healthy person to feel that life simply is not worthwhile.

So, what is this awful thing we call depression?

How can we equip ourselves to fight against something totally outside our own experience?
Is it even necessary to have such an experience in order to help someone else?

As a doctor I am sure that I have a better understanding of anything of which I have had personal experience, whether it be an investigation technique, an operation or an illness, and medical students are encouraged to gain as much of this type of insight as possible.

The success of the AA (*Alcoholics Anonymous*) proves that an ex-alcoholic is the best person to help an alcoholic - yet the nature of life is such that one cannot possibly have personal knowledge of everything with which one has to deal, and certainly one does not have to be an ex-depressive to help someone suffering from depression.

Insight, however, is essential,

and insight can be achieved by an understanding of our psychological makeup.

On the wall of the ancient Greek Temple to Apollo in Delphi are written the words
"KNOW THYSELF"

What a beautiful, enigmatic and thought-provoking phrase!

One of its interpretations is that of Alexander Pope in his lines:

"Know then thyself, presume not God to scan,
The proper study of mankind is man."

Relating this to our subject, I suggest that the answer to my question "What is depression?" ultimately lies within ourselves however happy we may be.

Returning to ancient Greece, we often say - *"The Greeks had a word for it"*, and true to form they had one to fit this particular bill: MELAGKHOLIA, which slightly modified to 'melancholia,' has been used in the English language until quite recently as a direct substitute for depression, and is still used today in a slightly different context as melancholy.

Literally translated Melagkholia means 'Black Bile' and was one of the four 'Humours', the interplay of which was believed by the doctors of those times to control the moods of man.

'Blood', 'Phlegm', 'Choler' and 'Melancholer' were supposed to be responsible for passion, coolness, anger and depression respectively and, naïve as this may seem to us today, these ideas were believed in this country until quite recently.

The important thing is to look past the naivety and try to find the element of truth which exists in the ancient concept.

The first truth is that melancholy is a **natural part of life** and human experience, and there is nothing unusual about its appearance in the right context.

Nor is it necessarily unpleasant, indeed it can sometimes be thought to be desirable;

The 16th Century philosopher Robert Burton wrote a lengthy poem entitled *'The Anatomy of Melancholy'* in which two sharply contrasting descriptions occur;

*"If there be a hell on earth, it is to be
found in a melancholy man's heart."*

and later

*"All my joys to this are folly,
Naught so sweet as melancholy."*

The latter of these interpretations may not be applicable to what we call depression, but few would disagree that a reaction of depression to an appropriate stimulus can be **completely natural**; in which case we call it **grief,** and if this did NOT occur we would regard it as very strange and indeed undesirable.

The second truth in the Greek Concept is to regard mood as a balance of different emotions, although nowadays we recognize many more than their four 'humours'.

Every one of the many emotions we can experience tends to have an opposite emotion which acts like its mirror-image.

For example:

Depression	Elation
Diffidence	Confidence
Anxiety	Tranquillity
Anger	Peacefulness

Patience	Urgency
Tolerance	Prejudice
Love	Hate

And many more.

We can picture personality diagrammatically as a circle with each emotion represented by a quadrant of variable size.
When two opposites are of equal size they balance each other producing an equilibrium.

In a normal, relaxed frame of mind, all of the quadrants tend to be balanced out, each being of approximately the same size as its counterpart and no emotion is dominant.

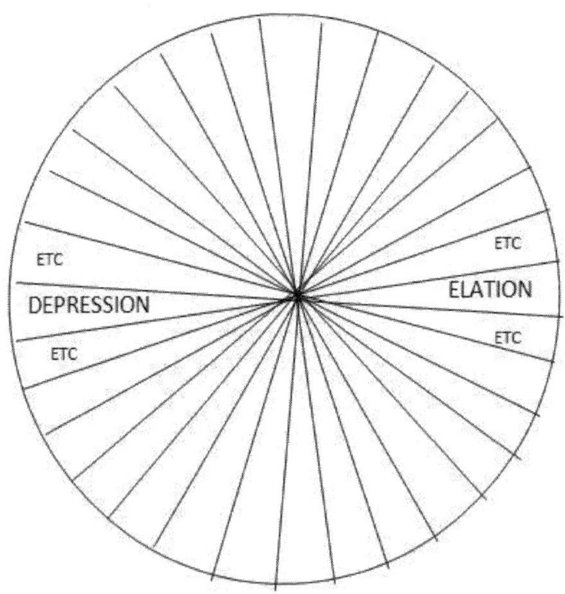

When an emotion is stimulated, its quadrant expands, firstly at the expense of its opposite number, which shrinks down proportionately and subsequently, if it intensifies, all the other quadrants begin to shrink down to accommodate it.

In the heat of battle a soldier may, at the time, be completely unaware of a serious injury and athletes achieving a high level of performance can be aware of nothing other than their personal goal.

Similarly, in the case of depression this emotion can dominate everything else;

Where I think the Greeks got it wrong, is that they regarded mood as secondary to the proportions of their humours and that mood then produced appropriate thoughts.

I think that this puts the cart before the horse.

What really happens is that
THOUGHT PRECEDES MOOD.

This is easy to understand in the case of grief, or any other example of what is commonly called *'Reactive Depression'*, which means just what it says.

Furthermore, I find that I can quite easily relate it to everyday experience and that whatever mood I am in can be accounted for by the circumstances of my life at that time.

This can be noticed particularly on waking up in the morning; most days on 'coming to', my mood is a sort of 'middle course' contentment. Neither 'up' nor 'down'.

Sometimes I wake up feeling excited, and occasionally rather 'low', and can find myself asking

"Why do I feel like this?"

Immediately I remember the pleasant or unpleasant events awaiting me that day.

There is another interesting aspect of this awakening experience. For me, identifying a pleasant stimulus serves to enhance the pleasant mood, whilst identifying an unpleasant stimulus, almost always causes the 'sinking feeling' to assume its correct proportions.

If then we can accept that, in normal circumstances, thought precedes mood and that there is nothing surprising or wrong about a normal reaction produced by a normal stimulus then

WHEN AND WHY DOES DEPRESSION BECOME PATHOLOGICAL?

Imagine two parts of an engine held together by a bolt. The joint will fail **either** because the bolt is weaker than expected – too small, rusted or overtightened,

or because the force exerted on it becomes unreasonably great.

Now, imagine a stressful life-situation which gradually increases in strength. At some point it becomes big enough to trigger an emotional reaction.

Let's call this point THE MOOD THRESHOLD

In just the same way as a mechanical bolt snapping, reactions (mood changes) can occur in two areas:

AREA 1   (compare a weak bolt) The position of the mood threshold is shifted making it more sensitive,

AREA 2   The stimulating force is exaggerated or amplified.

AREA 1 <u>Examples of are</u>:

<u>Mental illness</u>. Commonly called 'Endogenous Depression'.

<u>Hormone Changes</u>. e.g. premenstrual and post-partum changes in women and 'change of life' in both sexes.

<u>Some Drugs</u>

<u>Long term stress</u> e.g. frustration in an obsessional personality, chronic illness, grief, anger, sexual frustration, cold, hunger, financial insecurity.

Apart from perhaps helping to ease stress, most of the above are the province of the medical profession.

AREA 2 *however has great potential for self-help, and the active participation of relatives and friends.*

Stimulating forces become exaggerated by any way in which the true facts are misinterpreted, i.e.

reality becomes distorted.

When this happens, the stimulus is no longer a 'normal stimulus' and the mood threshold, *wherever it may be,* is crossed more easily.

**As soon as the mood threshold is crossed
and the mood of depression becomes
established a new set of circumstances exists.**

- There is a further increased sensitivity of the mood threshold until it becomes 'hair trigger'.

- There will be an increase in thought distortion consequent on the depressed mood.

In other words, **a 'vicious circle' has been established** from whose grip it becomes increasingly more difficult to escape.

Furthermore, the deeper the sufferer goes the more difficult it becomes for them to have an insight into their own condition.

See the 'depression awareness' quadrant described on page 9, which expands until it occupies the whole of their consciousness and is in danger of overwhelming everything else.

It is essential to break this downward spiral.

Identifying distorted reality is where friends and relatives, who have an insight into what is happening, can be of enormous help, even though they may be unable to influence the primary shift of the mood threshold.

Abnormal thoughts can be identified, and the sufferer can be helped to recognize them themselves, to counteract them and eventually learn not only how to:

prevent crossing the threshold into emotional reaction,

but also, how to

break the vicious circle once the threshold is crossed.

DISTORTION OF REALITY

Understanding of what this means is crucial.

I'm not talking about hallucinations, dreams or delusions but ways in which all of us can sometimes 'get things completely out of proportion.'

> When under stress, normal logic and common sense fly out of the window.
> Often these distortions can be extremely persuasive and difficult to recognize.

> The important thing is-

To know exactly what one is looking for.

It can be surprising how often a long-established distortion can suddenly stand out in sharp relief when the right sort of light is shone upon it.

I think it is helpful to form the habit of looking for distorted thinking under four headings:

- Negativity
- Guilt
- Inferiority
- Emotional Flatness

These divisions are not all that distinct, and distorted thoughts often overlap two or more types, or interrelate one with another.

The important thing is that most distortions fit in somewhere, and if you can remember to look for each one of these groups routinely, you won't miss a lot.

One of the features (not causes) of depression is a difficulty in concentrating, and therefore, to achieve maximum benefit, it is important **to keep self-therapy as simple as possible**.

Let's have a look at some examples of the four types of distorted thinking.

1. NEGATIVITY

This is PESSIMISM in capital letters.

Nothing good can be seen in the future and even when nice things happen they are ignored or cynically explained away.

"The future contains nothing but disasters."
"There is no point in trying."
"I will never be any better."
"Why bother to wash, shave, put on makeup, clean the house, or do anything properly - it doesn't help."

One failure or set-back means "It will always be the same."

Successes are forgotten or ignored.

"The sun may be shining today, but we will pay for it later."
"Things are all black and white – mostly black."

Holidays or outings are deliberately avoided because it is unbearable to be surrounded by people who are enjoying themselves.

If someone appears to like you it is only because they are "Being nice" or "Trying to curry favor."

"Why has this happened to me?"
"The world is not a fit place to bring my children up in – they would be better off dead."

2. GUILT

"It's all my fault."
"I'm upsetting everyone - the whole family – they try so hard to help and this is how I repay them."

Genuine or imagined past misdemeanors are blamed for present circumstances, mishaps or illnesses.

"I am being punished."
"I feel guilty, therefore I must BE guilty."

"I know I ought to do something to help myself, but I just can't get around to it."   - for example, dieting if overweight.

(Sometimes the latter can lead on to going in the opposite direction in the form of compulsive eating, which leads to more guilt.)

"My family would be better off if I were dead."
"I don't deserve help."

3. INFERIORITY

"I'm a failure." (Even when quite the opposite is obvious).
"I'm no good to myself or anyone else."
"I'm ugly."

Weaknesses or imperfections are exaggerated whilst good points are minimized.

A student may get 6 grade A passes but conclude that they have 'failed' because the 7th was "only a B."

"No-one loves me."
"I'm not worth helping anyway."
"Tom, Dick and Harriett are all more successful / cleverer / happier than me."

One of the most successful comedians of my youth was Tony Hancock, whose comedy skills were admired by everybody but himself. Writing about him with the sympathy and understanding of another depressive, Spike Milligan tells how his feelings of insecurity, unworthiness and self-doubt recurred between each performance. He reacted by firing one after another of his comedy team – Bill Kerr, Kenneth Williams, Hattie Jacques and Syd James. When they were gone he fired his script writers Ray Galton and Alan Simpson.

Finally, there was no-one else to get rid of but himself.

4. EMOTIONAL FLATNESS

"I am incapable of loving or being loved."

"I am more like a robot than a human being."

"I'm impotent, can't concentrate, can't remember."

"I used to laugh a lot but now I seem to have forgotten how."

"Do what you want with me – take me away, lock me up.
I don't care anymore."

Depression is often accompanied by painful self-harm and suicide attempts are frequently of an extremely painful nature such as jumping from great heights or under trains or cutting the throat.

DEALING WITH DISTORTED THINKING

- **RECOGNISE IT.**
- **RATIONALISE IT.**
- **GET IT IN PERPECTIVE.**

If they are willing to try, friends or relatives can be an enormous help to the sufferer in achieving these objectives.

The most important thing to do is to LISTEN.
The next is to be PATIENT and UNDERSTANDING.

Remember that one of the features of depression is difficulty in concentration. It may be helpful to, together, make a list of any 'bad thoughts' and then analyze them individually to see if they fit in with any of the four groups, negativity, guilt, inferiority and emotional flatness. If they do, then they should be seriously assessed for truth or distortion and a positive search be made for compensating factors.

The next step is to help to replace distorted thinking with positive, rational, accurate thinking. This is really re-learning the habit of 'looking ahead' or 'looking forward', which is second nature to most of us, but which tends to be completely lost in depression. This is not as easy as it might sound but can be helped by **adopting a specific technique**. This is simple, (if not easy) but requires sustained effort.

Once again, gentle encouragement can be enormously helpful.

THE ART OF FORWARD THINKING

First let's consider how our subconscious normally functions.

Contrary to the words of the old song, in reality,
life is NOT just a bowl of cherries!

The truth is that the majority of it is HUMDRUM with quite short interludes of pleasure or displeasure.

Imagine that we can measure pleasure and plot it on a graph (up and down) against time (from left to right)

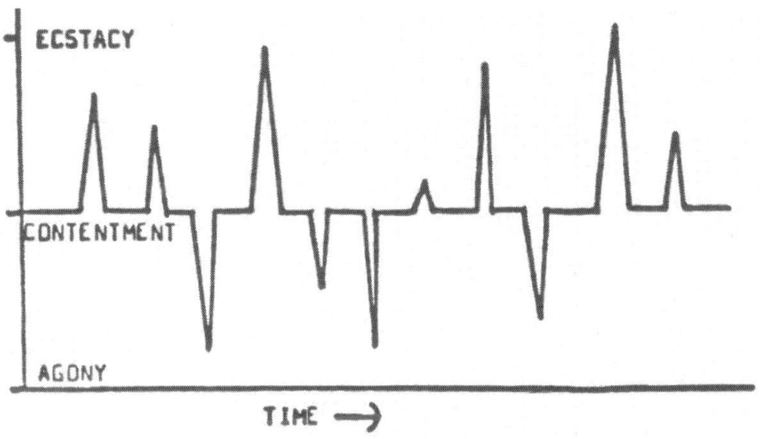

For most of us, for the majority of our day, the result would be a fairly straight line, midway up – 'Contentment'

Interludes of 'pleasure' and 'displeasure' would be represented by upward or downward 'spikes.'

21

Although the pleasure or displeasure can be quite intense, producing a large deviation from the normal, because they don't last very long they can be quite narrow at the base.

The sum-total (represented by the area occupied by the triangle) is therefore quite small however high or low it might be.

Consequently, the amounts of the feeling experienced are comparatively small when compared with the whole 24 hours of the day.

However, in normal circumstances we widen the base of the pleasure triangles by

Anticipation beforehand and
Savouring pleasantly afterwards.

The more we do this the greater the pleasure experienced.

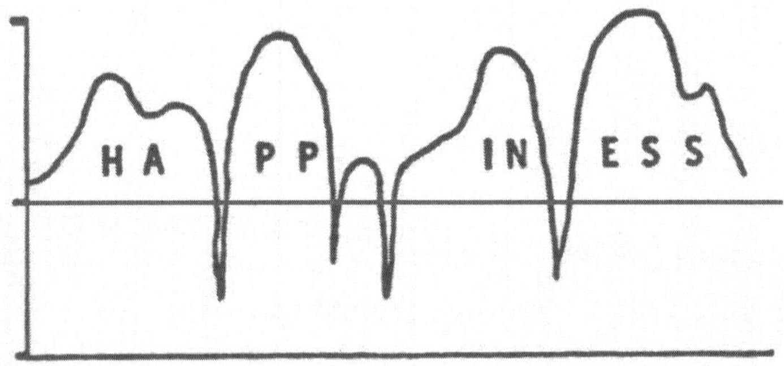

The depressed person still has the upward spikes, but they come and go so quickly that they are hardly noticed.

Conversely, they do exactly the same sort of thing to the downward spikes, often widening them out until they occupy the whole graph.

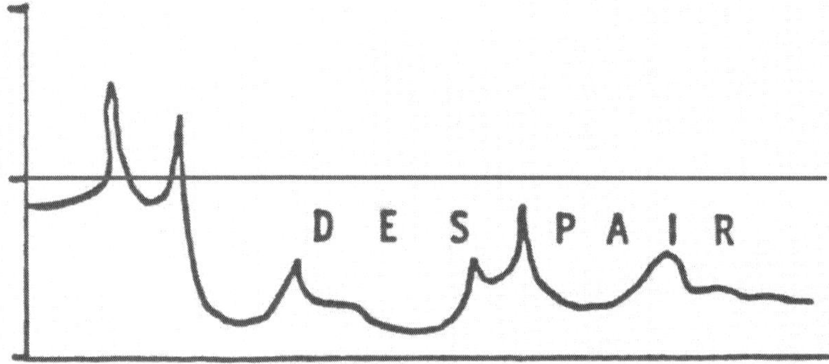

What is needed is help to re-acquire that normal habit of positive optimism, so that life becomes something to be looked-forward to and enjoyed.

ACQUIRING THE HABIT OF FORWARD PLANNING

Once again using a paper and pencil can be very helpful.

I try to encourage the habit of spending a few minutes each day planning ahead – usually just for the next few hours – later expanding to the whole day as the practice becomes more familiar.

No great detail is needed, just a skeleton time-table with a few basic ideas (high spots) as to how the time is to be spent.

When there is a big empty space in the time-table – and to begin with this will be most of the time – then an effort must be made to put SOMETHING into that space. **(Help needed).**

It doesn't really matter what it is, although it is better if it is something pleasurable.

At first this can be _very hard,_ which is when friendly encouragement and a bit of 'prodding' is invaluable.

- Look at radio and television programs to see
  if there is anything to interest you.
- Get interested in a book (a little at a time).
- Plan to visit a friend.
- Arrange a small treat for yourself – a tea-break.
- Try to get interested in ongoing, creative things.
- Try to include something physical. Go for a walk.

Often regular routines will be established.

Then, on waking up tomorrow, instead of

"Oh hell, another day to be got through!

the day will have SHAPE and MEANING.

Morale can be boosted in lots of positive little ways.

These should be identified and then looked for and included in the planning process.

For example, we all have certain clothes that we enjoy wearing and others not quite so pleasing.
Deciding to wear the former is all part of positive planning.

Spending a little extra time on make-up, even when the lady has no-one to see it, is never wasted time.

It is beneficial to incorporate as many as possible of these morale boosters in the days plans.

Ask a depressed person what their hobbies are, and they will probably say "I haven't any", but some sort of regular activity apart from work is an extremely important part of life.

It's well worth delving a bit deeper, perhaps to discover some forgotten pastime or old interest which can be rekindled into renewed life, or new suggestions made which might be appealing.

Lists of local clubs and societies, including a wealth of subjects available at night school and U3A, can be obtained from Local Information Offices, where the extent of available activities can be quite an eye-opener.

The precise nature of a hobby doesn't really matter, so long as it is something which can be enjoyed for its own sake, taken up or put down at will and can fill some of those blank spaces.

The very best of interests are those that are **creative** in some way. This doesn't necessarily mean making something elaborate, but just that the activity has continuity, progressively working towards a conclusion, which when reached, can create a sense of fulfilment and an enthusiasm to go on to the next venture.

To begin with, something as simple as a jig-saw-puzzle can be most therapeutic, but ultimately the greater the potential for achievement, the greater the value and it soon becomes second nature to fill those blank spaces in the days program.

Life is a great adventure and filled with experiences, many of which in themselves are unpleasant, but if only these can be seen as part of one's total existence, and savored as such, then even the unpleasant things can be appreciated.

**Even a visit to the dentist is better
than having nothing to do.**

Think about it!

WHY DOES ENDOGENOUS DEPRESSION OCCUR?

Now that you, hopefully, have a better understanding of the nature of depression, I am going to digress for a little while to consider possible reasons for its occurrence.

I may be wrong, but it seems to me that **depression is, to some extent, a product of modern civilization.**

There certainly seems to be a progressive increase in the number of people seeking help for this condition.

If this perception is true, then it seems to be far more likely that the reason is to be found in the field of distorted thinking rather than in that of a primary shift in the mood threshold i.e. in what is classified as 'Endogenous Depression.'

An interesting fact is that, so I am told, **depression is virtually unknown amongst aborigines in Australia.**

Some people draw the conclusion that this is due to a lower level of basic intelligence, and Aborigines are thought by them to be lazy, feckless, dirty and primitive.

This is a complete misconception.

The truth is that they have an **entirely different philosophy of life,** and things about which we get 'up-tight', to them are of no consequence. Aboriginal babies brought up in what we regard as conventional surroundings develop IQs and intellectual abilities comparable with any other race.

Furthermore, in their natural surroundings, they are brilliant linguists. There are over 100 completely different 'abo' languages and thousands of dialects. The majority of aborigines are fluent in several of these languages, many in 10 or 12, plus any non-Australian languages with which they happen to have been in contact.

How many languages do you speak fluently?
So, what is it that makes them the way they are?

More importantly, does whatever it is, counteract the development of depression?

I don't know the answer to this, but it may be something to do with their attitude to life.

**Perhaps Modern Western Society has
got its priorities wrong!**

Is it perhaps something to do with the modern preoccupation with 'Keeping up with the Joneses' and the emphasis on material success?

This is quite capable of putting a critical strain on a susceptible personality, whatever their actual achievements may be.

**It is always possible to find a more successful
Mr. & Mrs. Jones!**

Don't get me wrong, I'm not decrying aspirations, which are an essential part of anyone's life, but sometimes people set themselves unrealistic targets which are totally impossible to achieve and then are unable to settle for anything less without feelings of inferiority and failure. This is not helped by the attitudes of 'Society': those 'in authority', teachers, bosses, ambitious parents and the media, who are constantly stressing the need to succeed. Sadly, the consequence of adversity in the form of unemployment, redundancy, business difficulties etc. can be construed as personal failure and inadequacy, even though the circumstances are completely beyond individual control.

It seems to me likely that the decline in formal religion has also played a part. A religious belief, whatever form it takes, is a powerful antidote to negative thinking. At the very least the sentiment "It's the will of Allah", takes some of the hopelessness out of life's circumstances and suggests some underlying reason, even if it is not understood.

Furthermore, most religions suggest that "God helps those who help themselves".

Less obviously, but probably more important is the question of guilt. The Confessional is a most effective 'safety valve' and even in religions where confession is not practiced formally, the belief in repentance and forgiveness usually far outweighs any feelings of unworthiness that may be produced.

[It should however be remembered that the deliberate provocation of feelings of religious guilt can sometimes be instigated by unscrupulous individuals to achieve their own ends.]

ANXIETY

It is absolutely natural for anybody who is ill in any way to feel anxious, particularly when they don't really understand the illness. In the case of someone suffering from depression, the tendency to distort reality sometimes involves the ability to handle everyday problems, and normal 'worry' becomes exaggerated until it becomes a dominant feeling of anxiety.

In other words, the anxiety quadrant, previously described, becomes enlarged in tandem with the depression quadrant.

**The consequence is that
depression and anxiety tend to go hand in hand**

The converse of course also holds true in that unresolved anxiety, what we sometimes call an **Anxiety State** can frequently be a cause of depression.

In an Anxiety State the 'mood threshold' is shifted in the direction of increasing anxiety and any slight worry, which in normal circumstances would be dealt with routinely, becomes exaggerated into a big problem.

Once again, the sufferer becomes involve in a vicious circle - a downward spiral of anxiety and the worse it gets, the more difficult it becomes to break free.

In these circumstances the mood threshold for both depression and for anxiety can become 'hair trigger' which can prevent either from being easily dealt with without some plan of action.

COMBATING ANXIETY

Anxiety tends to be reinforced by avoiding making decisions.

The need for decision making is happening all the time and may concern;

- the consequences of 'outside' future events, from
  "Tomorrow's weather" to
  "The outbreak of World War III."

- the consequences of our own decisions from;
  "Which socks am I going to put on today?" to
  "Shall I change my job/partner/study-course" etc.

The problem with an anxiety state is that it becomes increasingly difficult to make any decisions at all.

We no sooner start to think about one problem, than another bursts into our consciousness, then another and another until we are bordering on a state of panic, so that

none of the problems ever gets dealt with.

Once again, a **plan of action** is required, and should be established well in advance of that downward spiral starting.

Once again it requires a pencil and paper.

Once again, a third party may be of great help.

THE STRESS-BUSTING PLAN

A. Make a list.
To begin with this is just:
A list of any and all the worries you have,
whether hugely important or trivial.

This may be difficult at first and your mind will keep wandering off in different directions usually to other worries, which

should then be added to the bottom of the list.

At this time you are just writing a list and
must avoid the desire to think any further about each item.

B. Start to consider the first item on the list.

No doubt immediately, other thoughts and problems will intrude into your consciousness

C. Add these new items to the bottom of the list but nothing more

Return to considering no 1 on the list.

D. Ask yourself the question.
"Can I influence the outcome of this topic, Yes or No?"

Answer is "No"	Answer is "Yes"
E. Take your pencil and score out item 1.	**Proceed to F**

F. Ask the question
"Can I influence it TODAY, Yes or No?"

Answer is "No"	Answer is "Yes"
G. Take your pencil and score out item 1.	**Transfer item 1 to a new 'To Do' list.**

H. Go back to the original list and repeat B to H.

The physical action of scoring out is very important as it has the effect of removing this particular worry from your subconscious.

"It has been dealt with."

At the end of the exercise you will be left with

- List number 1 which has all been scored out.

- List number 2 - 'Your action plan' for today.

All worries which are beyond your control either immediately or long term **have been dealt with.**

Your 'to do' list will have been reduced to a **very small** number of things that need to be sorted out at this moment, and you can get on with dealing with them either yourself of with the help of others.

SUMMARY FOR HELPERS

Depression in someone close to you can be a very frightening phenomenon which I hope will have become a little less daunting through an increased understanding of its nature.

The best advice I can give you is to remember that you have two ears and one mouth. In other words,

**the most important thing you can do
is to listen and empathize.**

Empathizing means compassionate sympathy, whilst not allowing yourself to become emotionally involved. This can be hard when the sufferer is very close to you.

**Your role is not to be a psychotherapist or councillor,
but a supporter, friend and confidante.**

I hope that having some basic structures will help you to have the confidence to adopt this role.

I have offered some suggestions as to how you can help to
- Identify and correct distorted thinking.
- Acquire the habit of foreword thinking.
- Deal with anxiety.

This methodology is designed to help you as much as the sufferer and should only be introduced very gently and slowly. Listen to what is said, which will <u>perhaps</u> give a lead into one or other of these techniques and they should very definitely **not be all introduced at the same time.**

AN INTERESTING TRUE STORY

Shortly after the end of World War II, a Mid-European psychiatrist came to this country. He had a colleague who had arrived a few years earlier and who had established a practice over here and was sitting in his waiting room until his friend was free.

A stranger entered the room and sat down, looking rather distressed. The psychiatrist smiled at him, whereat he burst into tears and after a while began to pour out his problems. Our friend listened to him for a lengthy period before the man became calm and after a further interval got up and left.

Some 10 years later the psychiatrist chanced to meet the man again, who thanked him profusely and told him that on their first encounter he had saved his life, as at the time he had been on the point of committing suicide had it not been for the psychiatrist's help.

At the time of that first encounter the psychiatrist spoke virtually no English but he was able to project sympathetic understanding and to LISTEN.

In conclusion, it is my earnest hope that these thoughts of mine can be the means by which some depressed people will be helped to recognize the truth about themselves and their surroundings, to seek out and eliminate their distorted thinking and to acquire the habit of positive thinking, thereby helping them to permanently break the vicious circle and

Allow them to re-join the human race.

Other books by Rick Appleton
[Available from Amazon.co.uk]

With Joy the Stars Perform their Shining

Combined with a fascinating travelogue and many interesting anecdotes this book aspires to establish a bridge across the gap existing between people of similar ideas and principles, but who are separated by different religions, denominations or ideology, and includes those who do not profess allegiance to any formal religious group.

Acknowledging the need for tolerance and humility it sets out to establish a common ground for all men and women of goodwill who are active in the pursuit of their ideals, irrespective of the fundamental individual beliefs with which they happen to have been brought up or which they have acquired through personal experience. It does not set out to challenge the opinions and beliefs of others but rather to seek out areas of agreement.

Apple Pie

'Apple pie' is an autobiography tending towards a mini saga as it includes memories covering a total of 160 years. It creates a fascinating insight into 20th Century middle-class English Life, much of which provides interesting comparison with that of the 21st Century.

The Mature Motorists Driving Coach

This book is not for Learner Drivers. It is for experienced 'Mature Motorists' who wish to maintain and even to improve their driving proficiency whilst not aspiring to taking a course in advanced driving. It represents a 'do it yourself' refresher course for all who pride themselves on being 'good drivers' and are concerned about road safety.

Rick is a National Observer with IAM RoadSmart (the Institute of Advanced Motorists).

Whither Away?

The funeral service contains the words;

"We blossom like a flower then wither away."

Has Mankind lost its Moral Compass?

Where are we Going?

'Whither Away' was written partly as a response to Richard Dawkins' best-seller 'The God Delusion', and partly as an attempt to make sense of religiously inspired atrocities.

It poses the question:

*"What is the place of religion
in a 21st Century scientific world ?"*

Made in the USA
Monee, IL
07 July 2026